1 NUTRIENTS

Chart 1 will help you.

Name ....

A Do You Know?
1. How many groups of nutrients are there? 1 2 3 4 ....
2. Name three of these groups. ....
3. Can you name the other groups? ....
   ....

B Fill in the Spaces
1. Nutrients are the f__d we ___ and the _____ we drink.
2. Nutrition is find___ out about the foods we dr___ and ___ and how ____ affect our h_____.
3. There is no one food that gives your b___ all the nutrients it _____.
4. ___ humans need food and drink.
5. Knowing about nutrition makes it ____ to _____ foods wisely.

C Complete this Chart

MY BODY NEEDS	
P_____	for _____ and _____
__ter	for _____ and _____
C_____	for _____
F___	for _____ and _____ and _____
__amins	for _____
F__re	for _____
Min_____	for _____

D Your Greatest Asset is your Health
Is this so? What do you think? Why do you think this so?

....
....
....
....
....

1 NUTRIENTS
Answers

B 1. food, eat, liquid
   2. finding, drink, eat, they, health
   3. body, needs
   4. All
   5. easy, choose

C Protein   growth   development
   water   cooling   transport
   Carbohydrate   energy
   Fats   warmth   protection   energy
   Vitamins   protection
   Fibre   digestion
   Minerals   regulating

2 PROTEIN

Chart 2 will help you.

Name ....

A Do You Know?

1. Animal protein is one kind of protein. What is the other kind?

   ....

2. Write down two reasons why your body needs protein.

   ....

3. How are plant and animal protein different?

   ....
   ....

4. Your body can store some nutrients. Can it store protein? ......

B Likes and Dislikes
(Do this by yourself and then in groups of four.)

1. Write down all the different sources of protein you can find.

....	....	....
....	....	....
....	....	....
....		

2. Tick the ones you like.
3. Draw a ring around the foods you have never eaten.
4. Draw a line under the foods you dislike.
5. Discuss your list with the group.
6. Record the three most popular sources of protein in your group.

   ....

7. Record the three most disliked foods in your group.

   ....

8. Record any foods that no one in the group has ever eaten.

   ....

9. Share your group's findings with the rest of the class.

2 PROTEIN
Answers

A 1. Plant protein
   2. Building new cells; repairing old cells
   3. Animal protein is complete. It contains the 8 essential amino acids. Plant protein does not contain these 8.
   4. No

3 CARBOHYDRATES

Chart 3 will help you.

Name ....

A Make a List

Worthwhile Carbohydrate-rich Foods	Non-Worthwhile Carbohydrate Foods
.... We call these worthwhile because	.... We call these non-worthwhile because ....

B Using Your Carbohydrate

- Activities that make us huff and puff use up a good deal of carbohydrate.
- Those that make us puff use up some carbohydrate.
- If we don't change our breathing pattern we are not using much carbohydrate.

1. How do you rate these activities?

   jumping over logs   reading a book   swimming a race
   walking fast   sleeping   digging a hole
   watching television   running fast   skipping
   riding in a car   riding a bike   walking up stairs fast

HUFF AND PUFF	PUFF	NORMAL BREATH
....	....	....

2. Which group of activities could you do easily for two hours? ......
3. Which activities could you do for only three minutes? ....
4. Tick the activities which you do.
5. Are you getting enough exercise to burn up the carbohydrate you eat? ....

4 QUICK QUESTIONS

Name ....

Charts 1, 2 and 3 will help you.

A Name two songs about food. ....
   ....

B Choosing foods wisely helps __ to develop and gr__ in a h_____ way.

C All foods are good for my body.
   Some foods are good for my body.   } Which is the true sentence?
   Most foods are good for my body.   Cross out the others.

D What Am I?

My 1st is in nice but not in ice.	. . . .
My 2nd is in munchy and lunch.	. . . .
My 3rd is single in but, but double in butter.	. . . .
My 4th begins radish and ends water.	. . . .
My 5th is in fish, spaghetti and vitamin.	. . . .
My 6th is my 3rd.	. . . .
My 7th is in oil and in calcium.	. . . .
My 8th is in potato, roast and boil.	. . . .
My 9th is my last and also my first.	. . . .

E Do You Know?
   1. What does the word 'protein' mean? ....
   2. My body is made up of .... of cells.
   3. What did Johnny Appleseed do? ....
   4. Who was the pavlova created for? ....
   5. Who was the peach melba created for? ....
   ....

F Find the Foods
   Look for the foods across, down and at a slant.

```
A B A N A N A P P L E G T K V M B E M A N D A R I N G F R U I T
A P R I C O T R I N A G L O B E A N T O R A N G E C H I C K E N
R I C E D R E M E E F I G L O A A B M R L T S U G A R S A L T W
S C R C A R R O T E M T O M A T O E L I N S T E A K U H O E S F
B E S A N D W I C H O P R G R I L L O N C O F F E E L D I L L T
```

4 QUICK QUESTIONS
Answers

B me, grow, healthy
D N
   U
   T
   R
   I
   T
   I
   O
   N
E 1. Of the first importance
   2. millions
   3. Planted appleseeds
   4. Anna Pavlova, the ballet dancer
   5. Dame Nellie Melba, the opera singer

5 VITAMINS

Name ....

Chart 4 will help you.

A A Very Short Quick Quiz
There are six groups of vitamins. Name them. ....

B Using the Chart
Choose one vitamin group, look at the chart, and answer these:
  1. Can your body make it? ....
  2. Can your body store it? ....
  3. Where does your body store it? ....
  4. How does it help your body? ....
  5. Which foods are rich in these vitamins? ....

C Make a Poster
Do this in groups of two.
  1. Choose a vitamin group (not the one you chose for Activity B) ...
  2. Collect this information about your vitamin group:

     Does your body make it? ....

     How does your body get it? ....

     Does your body store it? ....

     How does it help your body? ....

     If your body stores it, where does it store it? ....

  3. Make a big, colourful poster to show what you have found out.
  4. Talk to a group or the class about your vitamin poster.

D Right or Wrong?
  1. Your body needs small amounts of all vitamins every day. ......
  2. Your body cannot make vitamin A. ....
  3. A cat's body can make vitamin C. ....
  4. Vitamin B is only found in fruit and vegetables. ....
  5. The sun helps your body to make vitamin E. ....

Now change each wrong statement to make it right. (Cross out or add words.)

5 VITAMINS
Answers

D 1. right
     2. wrong — delete 'not'
     3. right
     4. wrong — delete 'only'
     5. wrong — 'does not help'

6 ASSIGNMENTS   Name ....

Charts 1, 2, 3 and 4 will help you.

Choose an assignment. Assignments take longer to do than other page work.

You could talk to other classes about your work when you have finished. Discuss this with your teacher first.

A *Tell it in Words*
Do this in groups of two or four.
We can describe food by:

| TASTE | SMELL | LOOK | FEEL | SOUND |

1. Write at least 10 words for each of these attributes.
2. Choose some of these words, and draw them in such a way that they suggest what they are.

B *Foods for Special Occasions, Places and Times*
Do this in groups of four.

• Special Occasion Foods	Birthday cake
• Special Place Foods	No hot roast dinner on a picnic
• Seasonal Foods	Salad in summer, soup in winter
• Time of Day Foods	No chocolate cake for breakfast

1. Choose one of these groups of foods. Collect all the information you can about it.
2. Make a decorative poster and present it to the class.

C *Staple Foods*
Do this in groups of two or four.
The main carbohydrate food of any nation is called its staple food. It is the main food in the diet. For example, Australia's staple food is wheat. Made into flour, we eat it in bread, spaghetti, etc.

1. Choose one of these nations and find out all you can about its staple food: China, Italy, Britain, United States of America, India, Japan, Mexico, Greece, Indonesia, Germany, Brazil.
2. Make a chart to record your findings. Include a map of the country and pictures and drawings of the foods. Give a talk to the class.
3. Discuss these questions:
   (a) What is the staple food of most people in the world? Why?
   (b) What is the staple food of most English-speaking nations? Why?

7 MINERALS

Chart 5 will help you.

Name ....

A Do You Know?

1. Fluoride is found only in _____. It is important for healthy t_____ and b_____.

2. Popeye always eats _____ because it is rich in _____.

3. C_____ and vitamin D work together to b_____ healthy bones and t_____.

4. Ph_____ helps muscles to work well.

5. Tomatoes, baked beans, bacon and fish all contain _____.

6. Minerals are as _____ for your body as all the other ___ major groups of nutrients.

B Where Do You Find It?

bananas		butter
margarine		spinach
nuts		corn
eggs	IRON	apricots
bacon	SODIUM	fruit
fish	POTASSIUM	apple
tomato	FLUORIDE	milk
salt	PHOSPHORUS	yoghurt
water	MAGNESIUM	chops
cocoa	IODINE	peas
baked beans	CALCIUM	cheese
dill pickles		liver
green vegetables		potatoes

C Unjumble the Sentences

1. body small in Your needs amounts minerals

   ....

2. includes a foods will A of which diet variety supply body with all your it minerals the needs ....

   ....

7 MINERALS
Answers

A 1. water, teeth, bones
   2. spinach, iron
   3. Calcium, build, teeth
   4. Phosphorus
   5. sodium
   6. important, six

8 FURTHER ASSIGNMENTS Name ....

Charts 1, 2, 3, 4 and 5 will help you.

Choose an assignment. You could do it by yourself or in a small group. Make sure that you plan your talk or poster carefully.

A Prepare a talk for another class about the importance of nutrition and choosing foods wisely. To make your talk more interesting, you could make a chart and leave it with the class.	E Collect pictures and labels of protein-rich foods. Sort them into animal and vegetable protein. Design a chart or book for these two sets of protein. Give it a title and label all the foods. Explain why protein is an important nutrient. Display your work.
B In our world, people die because they cannot get enough to eat. In our country, many people eat more than they need. What do you think about this? Prepare a talk.	F Make a bright picture poster to advertise worthwhile carbohydrate. Label the foods and explain why carbohydrate is important for the body. Display your work.
C In USA over $2 million each year is spent on advertising non-worthwhile sugar-rich foods. Survey the advertisements shown on TV between 4 pm and 6 pm, in local shops, in magazines and papers. What do you notice about the advertisements? Make a chart to show your findings. Present it to the class.	G It is important to care for your health. Make a poster to explain why this is so. Talk to the class about the poster.
	H Make an advertisement for radio or television about the many sources of vitamin-rich foods we can eat every day. Explain why vitamins are important.
D Make a poster to show why minerals are an important nutrient group. Use pictures or drawings to show your audience which foods are good sources of minerals. Display your work to the class.	I On Sheet 4 QUICK QUESTIONS, question F is a Find the Foods word game. Make another one and give a copy to everyone in the class.

9 FATS

Name ....

Chart 6 will help you.

A Question and Answer
Write a sentence to answer these questions.
Q. Why is fat an important nutrient?
A. ....
Q. Does your body use extra fat for fuel?
A. ....

Now write the QUESTIONS for these answers.
Q. ....
A. Vitamins A, D, E and K are found in some of the fats we eat.
Q. ....
A. Peanuts, chocolate and olives have what is called invisible fat.

B True of False?
1. Your kidneys, lungs, liver and heart have a special padding of fat around them. ....
2. Your body cannot store fat. ....
3. All foods contain fat. ....
4. Milk is a good source of fat. ....

C Finish these Pages

Foods that contain mainly unsaturated fat:	Foods that contain mainly saturated fat:
1 .... 5 .... 2 .... 6 .... 3 .... 7 .... 4 .... 8	1 .... 5 .... 2 .... 6 .... 3 .... 7 .... 4 .... 8 ....

sunflower oil   eggs   pork   milk
unsaturated margarine   veal   chicken   cheese
peanut oil   nuts   cream   beef
scallops   fish   olives   dripping

D What Do You Think?

1. Name eight kinds of fast take-away foods. ....
   ....
2. Why are they called the 'greasies'? ....
   ....

9 FATS
Answers

A   A   We need fat for warmth, extra energy and vitamins A, D, E, K.
    A   Yes, it does.
    Q   Which vitamins are found in the fats we eat?
    Q   Can you name three foods that have invisible fat?

B   1. true
    2. false
    3. false
    4. true

C   1 chicken   5 unsat. marg.   1 olives   5 cheese
    2 fish   6 sunfl. oil   2 cream   6 milk
    3 nuts   7 scallops   3 beef   7 eggs
    4 veal   8 peanut oil   4 dripping   8 pork

10 WATER

Name ....

Chart 7 will help you.

A *Finish these Signs*

```
Worthwhile Drinks
for Hot Days
• o___ge ___c_
• w____
• m_k
• i__ c____
```

```
Worthwhile Drinks
for Cold Days
• hot m___
• s_p
• hot _____te
• c____
```

B A fizzy drink is a non-worthwhile drink but milk is a worthwhile drink.

Why is this so? ....

....

....

C *True or False?*
1. I get some of the water my body needs from vegetables, fruit, meat, bread, cheese and cake. ....
2. I need to drink only one glass of water a day. ....
3. Humans can live without water. ....
4. Water helps keep my body at the right temperature: 59°C. ....
5. My body loses water through sweating. ....
6. Water transports the nutrients around the body. ....
7. Water is vital to my life. ....

D *Fill in the Spaces*

Water transports _____ nutrients around my _____ and it helps get rid of the w_____. Water also helps _____ my _____ at the right temperature: _____°C. My body uses about _____ litres of water each _____. On _____ days or when I _____ not well, I need to drink _____. More than _____ of my body weight is _____. I need to _____ about _____ glasses of water every day.

10 WATER
Answers

A
- orange juice
- water
- milk
- ice cream
- hot milk
- soup
- hot chocolate
- cocoa

B   Milk is a worthwhile drink because it also supplies the body with other nutrients such as fat, protein, minerals, vitamins and carbohydrate. Fizzy drinks contain just sugar and water.

C
1. true
2. false
3. false
4. false
5. true
6. true
7. true

D   the, body, wastes, keep, body, 37, ten, day, hot, am, more, half, water, drink, eight

11 FIBRE

Chart 8 will help you.

Name ....

A Complete the Chart

FIBRE-RICH FOODS			
a____	pe____	sp_____	to_____
cel___	___rot	p____ge	cu_____
cab____	pot___	ba____	cer___
le_____	br____	____kin	be___

B Which Fibre-rich Foods Would You Suggest?

SNACKS	LUNCH	BREAKFAST
....	....	....
....	....	....
....	....	....

C Unjumble these Sentences

1. found is foods that from plants. Fibre in come

   ....

2. rich like foods Fibre carrot help me also apple teeth clean. my keep and

   ....
   ....

3. body can't make fibre. My

   ....

4. every I ten fibre foods day. need eat to pieces of rich

   ....

5. essential is fibre good for digestion.

   ....

D Try These

- Freeze some orange quarters. Hey presto! A fibre-rich, icy, juicy snack.
- Make your own muesli for breakfast — plain or toasted.

11 FIBRE
Answers

A   apple   peach   spinach   tomato
    celery   carrot   porridge   cucumber
    cabbage   potato   banana   cereal
    lettuce   bread   pumpkin   beans

12 SCIENCE EXPERIMENTS Name ....

Charts 1, 2, 3, 4, 5, 6, 7 and 8 will help you.

A Finding Fat in Food

1. Put a drop of oil on a piece of paper. On another piece, smear a little margarine. Wait three minutes and you should have a grease spot. Now you know how to find out if there is fat in food.
2. Test other foods for fat — biscuit, apple, pastry. Make a prediction about the fat content before you test.
3. You could record your results like this:

FOOD	I THINK	I FOUND OUT
oil celery	yes no	yes

How do you get rid of grease spots on clothes?

B How Do We Know What We Eat?

1. You will need about ten small pieces of food, for example apple, salami, onion, cheese — and a blindfold.
2. Blindfold one person at a time. See if he or she can identify the food by smell and then by taste.
3. You could record the results like this:

JANE	SMELL	TASTE
carrot salami	no yes	yes yes

4. Which sense is more accurate: taste or smell? When? Why?

C Water in Food

1. Put about twelve small pieces of food on a sheet of paper in a cool spot. Leave them for seven to ten days.
2. Record what happens to the food.

DAYS	APPLE	BREAD	CHEESE
2 4 7 10			

What do you think?

13 CROSSWORD Name ....

ACROSS
1. Potatoes are _____.
6. A drink made from the leaves of a bush and boiling water
7. A sweet, juicy citrus fruit
10. Johnny Appleseed hoped everyone would have __ apple tree.
11. A fibre-rich breakfast cereal
13. Sometimes I'm visible, sometimes I'm invisible.
14. You ____ to know about nutrition.
16. In a song, I'm 'one a penny, two a penny'.
17. Pass the salt and pepper, plea__.
18. Lemo_s, _ranges and g_apefruit are citrus fruits.
19. All humans ne__ to eat and drink.
21. If you know about nutrition, eating well is _____.
22. Tomatoes grow __ bushes.

DOWN
1. _____ are an important nutrient group.
2. I hang in bunches from vines.
3. Carrots and turnips are ___ root vegetables.
4. I am a good source of plant protein. Jack planted me.
5. You can boil, fry, poach or scramble me.
8. I am found in buns and fruitcake. I come from dried grape.
9. During meals, people can be seen _____ and drinking.
12. I am a mineral.
13. I am necessary for digestion.
15. Snakelike fish that not everyone likes to eat
20. Eating is something that most of us like to __.

13 CROSSWORD
Answers

¹V	E	²G	³E	T	⁴A	B	L	⁵E	S
I		R		W		E		G	
⁶T	E	A		⁷O	⁸R	A	N	G	⁹E
A		P			¹⁰A	N			A
¹¹M	U	E	¹²S	L	I		¹³F	A	T
I			O		S		I		I
¹⁴N	¹⁵E	E	D		I		¹⁶B	U	N
¹⁷S	E		I		¹⁸N	O	R		G
	L		U			¹⁹E	²⁰D		
	²¹S	I	M	P	L	E		²²O	N

14 DIGESTION
Chart 9 will help you.

Name ....

A Label the Diagram

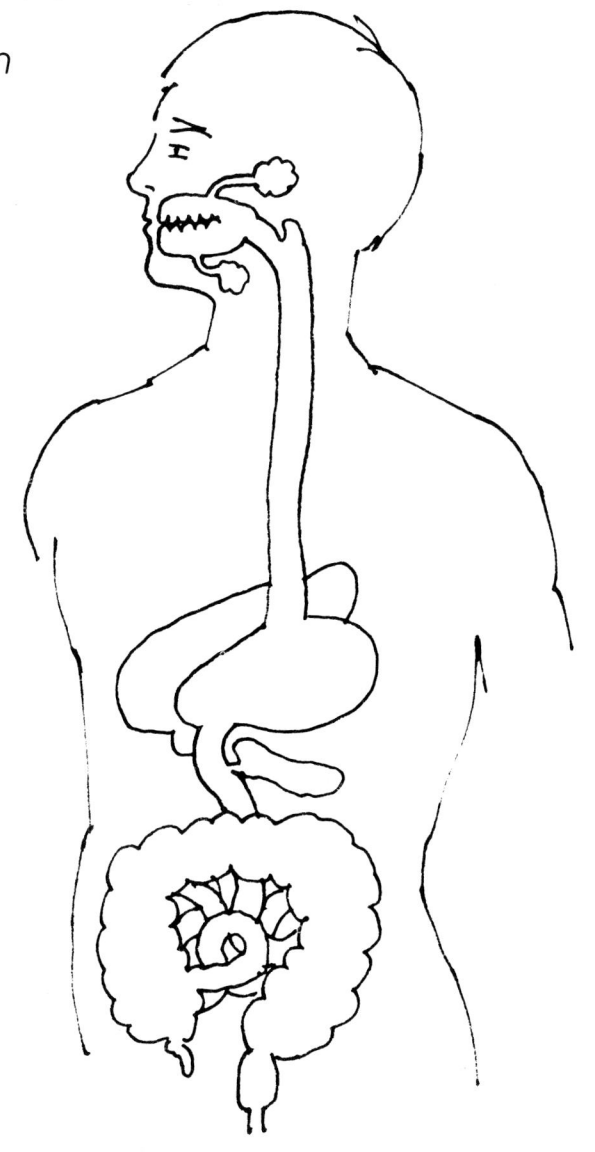

B Sentence Answer Quiz
1. Where do the nutrients pass into the body system?
   ....
2. Which nutrient helps the digestive process?
   ....
3. Why are your teeth important?
   ....
4. What carries the nutrients around the body?
   ....
5. What is the name of the special juice in your mouth that begins digestion? ....
6. Cows have two stomachs. How many do humans have?
   ....

14 DIGESTION
Answers

B 1. small intestine
    2. fibre
    3. begin to break the food up
    4. blood
    5. saliva
    6. one

15 THINKER PAGE   Name ....

A *Alphabet Puzzle*
   See if you can name a fruit that begins with each letter of the alphabet. On another piece of paper, you could try this with vegetables, or protein-rich foods, fibre-rich foods, and so on.

A	J	S ....
B	K	T ....
C	L	U ....
D	M	V ....
E	N	W ....
F	O	X ....
G	P	Y ....
H	Q	Z ....
I	R	

B *Make Sense of These Statements*
   1. My wonderful body chemistry can change _____ into vitamin A.
   2. Milk is a good source of ....
   3. An apple a day ....
   4. _____ is vital for digestion.
   5. We cannot live without _____.

C *What Am I?*

   I am rich in protein.
   I live in water.
   I have scales on my body.
   I am a ....

   I am rich in all nutrients.
   You eat me hot for breakfast.
   Bears and Goldilocks like me.
   I am ....

   I am rich in carbohydrates, vitamins, fibre, minerals, water.
   I have a stalk.
   I grow on a tree.
   I am ....

   I am a source of plant protein.
   I come in many varieties.
   You eat me raw, salted, roasted.
   I also supply saturated fat.
   I am ....

15   THINKER PAGE
Answers

B   1. carotene
    2. vitamins A & B, calcium, fat, protein, iodine
    3. keeps the doctor away
    4. Fibre
    5. water

C   fish   porridge
    an apple   a nut/peanut

16 MORE ASSIGNMENTS

Name ....

Charts 1 to 10 will help you.

A Make a chart to advertise good eating habits. Choose one of these topics:
- What's for Lunch?
- Great Breakfasts
- Delicious Drinks
- Super Snacks
- Terrific Dinners
- What's to Eat After School

When you have finished, present your work to the class or to another class. Display your chart.

An apple a day, keeps the doctor away.

C Find out how many different ways apples can be cooked and eaten. How would you cook and serve the following foods? Select some and illustrate the methods.

- carrots
- spinach
- potatoes
- sausages
- fish
- bananas

A garlic a day, keeps everyone away.

D Look at Chart 10 WHAT'S IN FOOD? You can make a chart like it. The other charts will help you. Cut out or draw pictures of about 12 pieces of food. Label each food with the nutrients it contains. Display your work.

B There are many proverbs and sayings about food. 'Too many cooks spoil the broth.' 'Ham actor.' Find as many as you can, and explain what they really mean. Share your work with the rest of the class.